BIOTECHNOLOGY DEVELOPMENT : THE ASIAN EXPERIENCE

Dr. Wagdy A Sawahel
Genetic Engineering and Biotechnology Division,
National Research Centre,
Dokki, Cairo, Egypt

1997

DAYA PUBLISHING HOUSE
Delhi -110035

ISBN 81-7035-181-2

Published by : **Daya Publishing House**
1123/74, Deva Ram Park
Tri Nagar, Delhi - 110035
Phone : 7103999

Laser Typesetting by : **Delite Printers**
Delhi - 110035

PRINTED IN INDIA

Preface

Biotechnology is the integration of biological and engineering sciences in order to enable the use of organisms, cells, enzymes and other derivatives (including molecular analogous) in industrial or commercial activities.

This monograph has been organized in three parts reflecting the industrial applications of biotechnology, the Asian experience in biotechnology development and it's implications for biotechnology communities.

Biotechnology could potentially affect any current industrial biological process or any process in which a biological catalyst could replace a chemical one. As discussed in part I, industrial applications of biotechnology will be found in several industrial sectors including: pharmaceuticals; animal and plant agriculture; speciality chemicals and food additives; environmental areas; commodity chemicals and energy production; and bioelectronics.

Biotechnology development depends on government funding of basic and applied research; personnel availability and training; University/Industry relationships; intellectual property law; health, safety and environmental regulations; financing and tax incentives for firms; targeting policies in biotechnology; international technology transfer, investment, and trade; antitrust law; and public perception, part II examines biotechnology development in some Asian Countries. It includes policy; traditional and new areas of biotechnology; legislation and intellectual property rights, commercial development; international relations; and future growth.

The Asian experience in biotechnology development is an ideal model. Thus, part III highlights the importance of such experience for science policy-makers, industrialists, trade officials and concerned citizenry.

Contents

I - BIOTECHNOLOGY : DEFINITION AND INDUSTRIAL APPLICATIONS

Biotechnology is broadly defined to include any technique that uses living organisms or parts of organisms to make or modify products, to improve plants or animals, or to develop microorganisms for specific use. It ranges from traditional biotechnology to the most advanced modern biotechnology. Commercial biotechnology consists of an expanding range of interrelated techniques, procedures and processes for practical applications in the health care, agricultural and industrial sectors. Commercialization of biotechnology ranges from research to products and services. These are powerful technologies, supported by complementary bioprocess engineering to help translate new discoveries of life sciences into practical products and services. As such, biotechnology should also be seen as an integration of the new techniques emerging from modern biotechnology with the well-established approaches of traditional biotechnology, such as plant breeding, food fermentation and composting.

In the past twenty years, dramatic new developments in the ability to select and manipulate genetic material have sparked unprecedented interest in the industrial uses of living organisms. Following the first successful directed insertion of foreign DNA in a host microorganism in 1973, researchers all over the world began to recognize the potential for directing the cellular machinery to develop new and improved products and processes in a wide diversity of industrial sectors. Potential industrial applications of those novel genetic techniques include the production of new drugs, food, chemicals, the degradation of toxic wastes, and the improvement of agricultural products. Thus, these new techniques could have a major economic impact on industries throughout the world.

Biotechnology has the technical breadth and depth to change the industrial community of the 21st Century because of its potential to produce substantially unlimited quantities of :

* products never before available,
* products that are currently in short supply,
* products that cost substantially less than products made by existing methods of production,
* products that are safer than those now available and,
* products made with raw materials that may be more plentiful and less expensive than those now used.

By virtue of its wide-reaching potential applications, biotechnology lies close to the centre of many of the world's major problems - malnutrition, disease, energy availability and cost, and pollution.

Industrial applications of biotechnology will be found in the following industrial sectors :

(A) Pharmaceuticals

As research using the new genetic techniques has progressed, the pharmaceutical industry has been the leader in industrial applications. Pharmaceutical production may be improved with biotechnology in many ways. In some instances, production of pharmaceutical products by chemical synthesis or tissue extraction methods may be replaced by production from cloned genes. In other instances, applications of recombinant DNA technology may supplant traditional bioprocess methods for the production of antibiotics and other pharmaceutical compounds. Perhaps most importantly, new biotechnology provides a means of producing - for the first time - amounts of compounds that are otherwise scarce. Thus, biotechnology may give rise to the development of entirely new pharmaceutical products among the most intriguing areas of research using biotechnology are those pertaining to the nervous system, and cancer. As research in these areas yields insight into mechanisms of disease and healthy body function, basic questions about the organization and function of the brain, the nature of behaviour, and the regulation of body functions may be answered. The illumination of these phenomena, in turn, may generate new possibilities for pharmaceutical products.

(B) Agriculture

As the world population grows, agriculture will need to provide more and more food. Biotechnology produced methods and products that improve agriculture in many ways. In animal agriculture, biotechnology offers promise in the following area :

* diagnosis, prevention, and control of animal diseases with the use of monoclonal antibody technology to diagnose, monitor, and better understand disease and the use of recombinant DNA to expand the pharmacopoeia of vaccines and other animal health products;
* animal nutrition and growth promotion through the use of growth hormones and feed additives to improve animal feed usage and animal health; and
* genetic improvement of animal breeds to better understand the bases of animal productivity and disease resistance and by the direct transfer of "beneficial" genes from one animal breed to another.

In addition, there are several potential applications in the area of crop improvement. The potential applications generally fall into two categories :

* improvement of specific plant characteristics, for example, through the introduction or manipulation of genes that confer resistance to disease and environmental factors, that increase the amount and quality of primary and secondary products from plants, that enhance plant growth rate, or that increase photosynthetic efficiency; and
* genetic manipulation of micro-organisms, for example, to enhance the process of nitrogen fixation, to produce insecticides, or to suppress disease or promote growth in plants.

(C) Speciality chemicals and food additives

There are many potential applications of biotechnology for the production of speciality chemicals. The nearest term applications are in the production of speciality chemicals that are produced by processes using micro-organisms, *e.g.* amino acids and enzymes. Enzymes are the direct products of genes, so their production is particularly accessible with new genetic technologies.

Current chemical synthesis production processes often require large energy inputs, have complicated synthesis steps, and yield many byproducts. If appropriate microbial enzymes (or higher organism enzymes) were identified and characterized, the appropriate genetic information could be cloned and expressed fairly rapidly in well-studied micro-organisms to produce or modify compounds such as vitamins, lipids, steroids, and aromatic chemicals.

The speciality chemicals and food additives market is extremely broad, and many other applications of biotechnology to the production of such products may be evident in the future.

(D) Environmental applications

Microorganisms have several uses in the environments and biotechnology can potentially be used to improve these microorganisms. One application is in the control of pollution and treatment of toxic wastes. Microorganisms are currently used in pollution control, and the potential applications of biotechnology to treat liquid and solid wastes are numerous. Additionally, techniques are beginning to be used to select microorganisms that can degrade extremely toxic compounds. In the mining industry, microbes are used to leach metals from mine dumps and concentrate metals from dilute solutions, and there are possibilities for using biotechnology to improve the efficiencies of these processes. A third environmental application of biotechnology is in enhanced Oil recovery.

None of the environmental applications of biotechnology are ready to be marketed, and there are still many technological problems to be overcome. Nevertheless, several companies are pursuing research and development in these environmental applications, and their development will progress over the next several years.

(E) Commodity chemicals and energy production

The Production of low-value-added, high-volume commodity chemicals demands the use of the most economic production schemes available. At present, the most economic schemes for chemical and energy production favor the use of petrochemical foodstocks. In the future, however, decreasing petroleum supplies, increasing oil prices, and technological advances in biomass utilization may foster a transition to the use of feedstocks derived from biomass. Such a transition is not expected to occur on an industry - wide scale in the near future, but bioprocesses are being used economically to produce significant amounts of fuel-grade ethanol from corn and other crops.

(F) Bioelectronics

The potential for the use of proteins in electronic devices have received attention recently with the advent of recombinant DNA technology and the potential for computer-aided design of proteins. Work is focused in two areas: biosensors and biochips. Biosensors (biologically based sensors) have been used for several years, but design problems have limited their acceptance. Biotechnology is expected to increase the variety, stability, and sensitivity of these devices. Biochips (biologically based microchips) capable of logic and memory are still only speculative, and their development is many years away.

II - BIOTECHNOLOGY DEVELOPMENT IN ASIA

The significance of biotechnology for Asian Countries could be illustrated by the example of Asian agriculture. In the early 1990s, Asia contained over 50% of the global population, over 70% of the world's farming families, but only 25% of the world's arable land. At the beginning of the 21st century, the per capita land availability would be 0.1 hectare in China and 0.14 hectare in India. The average Asian population growth was 1.86%. The only pathway open to countries like China and India for feeding their growing human populations was continuous improvement in yields. China had gone in for large-scale exploitation of rice hybrids for this purpose. Biotechnologies could help raise the productivity of major crop species.

According to the Food and Agriculture Organisation of the United Nations (FAO), Asian countries could be subdivided into the following four categories, depending on their degree of involvement in biotechnology:

Category I: Countries with an interest but no direct involvement in modern biotechnologies as yet, *e.g.*, Bangladesh, Bhutan, Burma, Nepal, Sri Lanka and Viet Nam.

Category II: Countries with a national policy and research programme, mainly in conventional biotechnologies, and which were monitoring development overseas but with little modern, in-country biotechnologies as yet, *e.g.*, Indonesia, Malaysia, Pakistan, Philippines and Thailand.

Category III: Countries with a national policy and a research programme, mainly in conventional biotechnologies, and which had established collaborative linkages in industrialized countries for training of scientists and acquisition of new technologies, *e.g.*, India, China.

Category IV : Countries with a national policy and research programme in modern biotechnologies, complemented by strong overseas linkage in both the public and private sectors, *e.g.*, Japan, Republic of Korea.

(1) CHINA

Biotechnology reached China in the early 1980's. The Chinese government is well aware of the value of high technology achievements in accelerating the industrialization of the national economy and biotechnology is considered as the first item among the eight fields of high technology which are being stressed. Several measures have been taken in order to facilitate developments in this area. In 1983 a Nation Center for Development of Biotechnology under the State Commission of Science and Technology was established as a leading and coordinating organisation.

The main objectives of the center are to formulate policy and programmes for the development of biotechnology, to organize, priority research projects, to coordinate research work of different institutions and to organize international exchange and cooperation. In 1984 a Chinese Consultative Committee for Biotechnology was set up. The task of this committee is to discuss and advise on the research and development of biotechnology.

The biotechnology programme consists of three levels in China. These are 1. the spark programme, 2. the 7th five-year plan (1986-1990) and 3. the high technology programme. Activities are allotted within these according to the complexity involved and basic work required. The spark programme consists mainly of the existing and easily adoptable techniques. The 7th five-year plan identifies the priority areas of scientific and technical development which have important economic benefit as well as some key technologies of strategic importance. The high technology programme is directed to more advanced technology and targeted at the most recent world wide developments.

The main research and development areas of biotechnology are: 1. breeding for high yielding, superior quality, and stress-resistant cultivars and more productive animals and fresh water fish; 2. developing new vaccines, medicines, diagnostic reagents and gene therapy techniques; 3. reforming the traditional biotechnology industries.

China is an agricultural country with 22% of the population of the world and only 7% of the land area of which 55% is arid or semi-arid land. Chinese agriculture is characterised by limited resources. Hence, biotechnology plays a particularly important role in attempts to maximize output at low cost.

The priority projects are breeding for crops resistant to pests and tolerant to salt and drought as well as screening hybrid strains with high yield, superior quality and other desirable traits. The improvement of the efficiency of nitrogen fixation and photosynthesis is considered as a long term target.

Development of new vaccines, medicines and diagnostic reagents using recombinant and hybridoma techniques have been active areas since the 1970s in China. Recombinant interferon and some monoclonal antibody diagnostic reagents have been marketed. Several recombinant vaccines and medicines are still in the pilot scale.

In the 1950's the fermentation industry changed from simple workshop production to a modern industry. Now the traditional industries, including antibiotic, amino acid, organic acid, beverage, enzyme, solvent, vitamin and steroid industries are well established. The market value of these products represent about 98% of the total output of the biotechnology industry. However, there is a need to reform the traditional biotechnological industry by reducing cost, energy consumption and pollution and to change the unreasonable production structure.

Efforts to increase penicillin acylase yield by using recombinant DNA techniques has met with success, this will promote the development of semisynthetic B-lactim antibiotics and improve the product range of the antibiotic industry.

Prior to the 1970s there had been some university or college departments and institutes which specialised in activities related to biotechnology such as fermentation, antibiotic production, biochemical engineering. In the 1980s the number of biotechnology departments and specialised courses increased rapidly to meet the perceived need so that there are now more than 50 universities, and more than 200 institutes involved in various branches of biotechnology. In these universities and institutes, most of the people have doctoral and master training and in some universities and institutes postdoctoral programmes have been established.

In order to bridge the gap between academic research and industry and to promote the commercialization of the new biotechnology (particularly genetic engineering and hybridoma techniques) the State Planning Commission established a new institute, The Shanghai Center of Biotechnology of the Chinese Academy of Sciences.

The Government has paid particular attention to regulations and guidelines for biotechnology relating to the release of recombinant organisms and the possible environmental hazard. This issue has been discussed in detail and relevant regulations and guidelines are being drafted.

International cooperation as well as export and import of techniques in biotechnology are encouraged by the government. Some production systems have been imported, for example techniques for single cell protein, restriction enzymes, cephalosporin, penicillin, embryo transplantation, hepatitis B surface antigen and themostable amylase production.

Meanwhile some biotechnological techniques have been exported, such as the processes for the production of monosodium glutamate, lysine, citric acid and vitamin C. The International Center of Genetic Engineering and Biotechnology (ICGEB) has approved the establishment of a subordinate center in China and has agreed to let Chinese scientists work at ICGEB institutes. An agreement on the establishment of a China-European Community Center of Biotechnology has been carried out between China and France, Federal Germany, United Kingdom, Italy and Greece. In addition, some cooperation agreements have been set up with the Rockefeller Foundation and other American, Japanese and Swiss institutes or companies.

(2) HONG KONG

The fate of Hong Kong in 1997 is one of today's true mysteries. However, many public and private sector sources quickly provide assurances that Hong Kong will remain a vibrant commerce center following China's takeover. It stands to reason that China can only desire such an outcome, as Hong Kong stands to be one of that nation's largest revenue producers, with its developed ports, strong international markets, and aggressive approach to business. In the area of biotechnology, Hong Kong is already home to several public and private research facilities—the most notable being the Hong Kong Institute of Biotechnology. When the institute was founded in 1988, it received $22 million in funding from the royal Hong Kong Jockey Club with the charter of promoting biotechnology in Hong Kong. Its goal has been to provide downstream support to basic biotechnology and serve as an incubator for the commercialization of biotechnology innovations. Located at the Chinese University of Hong Kong, the institute houses 75,000 square feet of administrative and laboratory space, and is in the process of developing a multipurpose bioprocessing facility. The Hong Kong Institute of Biotechnology (HKIB) is currently installing a 12,000 square feet bioprocessing unit that will include state-of-the-art microbial fermentation and mammalian, insect, and plant cell culture pilot plants, a downstream purification suite and several research labs. The bio-processing unit will focus on providing services to companies and universities that lack in house bioprocessing capabilities and foreign companies interested in assessing the Asia-Pacific market for products in its development pipeline. The HKIB is also involved in qualification efforts for certification under the World Health Organization's good manufacturing practice requirements.

Some examples of joint development efforts underway of the HKIB include the manufacturing of a recombinant malaria vaccine in conjunction with the US National Institutes of Health (Bethesda, MD) and in conjunction with US-based. Neuromedical Systems, Inc. (Suffer, NY), which is developing an image based tool for improving diagnosis of cervical cancer, and the radiopharmaceuticals Unit Dose Dispensing Center for the diagnosis and treatment of diseases with radioactivity, in conjunction with Amersham Radiopharmaceuticals (Buckinghamshire, UK).

Recently, the HKIB has also initiated an ambitious project to develop a serum-free culture medium formulation with applications that could range from the development of hormone therapy, growth factors, interferon, and monoclonal antibodies. This serum-free approach will provide substantial cost savings when compared to serum-supplemented counterparts. Other technologies in development include increasing the yield for certain plants and the development of ginseng-based foods and beverages.

Another notable research center in Hong Kong is the Biotechnology Research Institute located at city's University of Science and Technology. Its areas of research include drug delivery, Protein design, genetic engineering, biomedical instrumentation and diagnostics development, and Chinese herbal medicine.

Several international pharmaceutical and medical device companies have set up facilities in Hong Kong, including 3M, Abbott Labs, Conner Pharmaceuticals, Ferring Pharmaceuticals, Schering Plough, and Vickmans Laboratories.

(3) INDONESIA

Indonesia considers biotechnology as an important tools to be applied as a means to meet some of the national needs and thus considers such development as national priority. A National Board of Biotechnology, under the Ministry of Research and Technology, has been set up to define the priorities reflecting political willingness of the government to develop biotechnology in relation to many national programmes (health care, environmental control, food and agricultural industries) which can be accelerated in their development whilst also using natural resources efficiently. The first priorities lie in the medical and agricultural sectors as well in the establishment of bioindustries for using or treating agricultural wastes.

Industrial biotechnology: Food fermentation in Indonesia has been transferred from generation to generation in many villages where the traditional techniques are still used. Recently, the Government has set up improvement programmes for fermented foods in recognition of their importance in providing nutrients whilst removing many undesired toxic factors. A number of the old technological process are now upgraded. Currently medium - scale bioindustries are producing alcohol, beer, citric acid, monosodium glutamate, liquid sugar, single cell protein, soy sauce, and biogas.

Biotechnology and tissue culture: Indonesia uses this technique to develop plants that are less susceptible to bad environmental conditions and also for the production of uniform plants materials for export, industrial plantations, agrotourisms, and nucleus grains.

Biological nitrogen fixation: Since 1978, Gajah Mada University in Yogyakarta (Central Java) has produced two tons per week of a rhizobium peat-based inoculant for soybean and peanut production which is distributed to farmers by the Government. In 1981 a national programme on biological nitrogen fixation was established under the Indonesia Institute of Sciences. Research aimed at developing improved inoculants for other legumes is being carried out at Gajah Mada University, Bogor Agricultural University, and at the National Biological Institute (now Centre for Biological Research) in Bogor.

Medical biotechnology: The Centre for Medical Biotechnology Research and Development has been recognised since 1986 at University of Indonesia, Jakarta. The main activities are to make diagnostic kits for detection of important infectious diseases which become public health problems in Indonesia, such as typhoid fever, filariasis, hepatitis B and dengue hemorrhagic fever. A gene probe to detect *Salmonella typhosa* has been proven to be an effective and efficient method of diagnosis when combined with a biological-labeling system. Further research is currently being carried out on other bacterial and virus infections, such as enterotoxigenic *E coli* and Human Papilloma Virus. The gene probe for filariasis which was developed in Boston (US) has been used in endemic areas in Indonesia successfully. The genetic characteristic of Hepatitis B virus from many areas in Indonesia has been studied and sequenced since there are many subtypes in Indonesia which are important epidomiologically. Dengue hemorrhagic fever research has been aimed at understanding the genetic characteristics and finding immunogenic fragments so as to produce a reliable immunodiagnostic based on MABs. *In vitro fertilization* techniques have been applied in Indonesia since 1987, and by 1990 at least five babies had been born, with numbers increasing as more centres of reproduction are established.

Animal breeding: For cattle and pigs, *in vitro* fertilization techniques and embryo transfer are being used. At the Institute for Agriculture, Bogor, recombinant DNA techniques are being used to improve the quality of embryos in terms of physiology and nutritional aspects. Other groups at the same Institute are working on the improvement of fishery products such as shrimp and oyster.

The volume of agricultural waste generated annually by the agricultural sector is approximately 3-5 times that of the product itself. Studies concerning utilization of these wastes are carried out at Universities and Inter-University Centres in Bandung, Yogyakarta and Bogor, at Research Institutes in Bogor, Jakarta, Malang, Semarang, Yogyakarta and Bandung. A number of private sectors and state enterprises are cooperating with Universities and Governmental Research Institutes. Many have joint research with foreign counterparts. The Government plans to set up a Centre for biotechnical R&D at Cibinong, between Jakarta and Bogor. At Serpong, south of Jakarta, another training centre will be enlarged for training of manpower extending existing small scale activities on industrial and agricultural biotechnology.

To meet national needs for manpower for R & D in biotechnology, the government has set up an overseas training programme which enables young scientists to be trained for Biotechnology in Bandung, Yogyakarta and Bogor regularly run courses for young scientists or lectures to improve their skill and knowledge. The courses are open to participation by people from other Universities as well as the private sectors. Some courses have been conducted in cooperation with UNESCO, WJO, ANBS, Asian Development Bank and other donors.

(4) JAPAN

The report entitled "Research and Development Plan for Leading and Basic Technology in the Life Sciences" submitted by the Council for Science and Technology was approved by the Prime Minister of Japan in August, 1984. This is a promotional report which outlines current areas and targets of research and development which have to be emphasised in order to comprehensively, systematically and effectively promote the development of technologies such as recombinant DNA and cell fusion that are necessary for manipulating the genetic information system of living organisms. The basic objectives of biotechnology in Japan includes promotion of basic research and fostering creative, original research in the long-term. Strengthening of policies related to the organisation and funding of R & D as well as training of personnel and international exchanges.

Current targets and areas of R & D which are being stressed include: Systematic utilization of genes, cells and proteins; Identification and synthesis; Introduction of genetic material into cell with expression of genetic information; Common basic technologies.

These areas of interest reflect the basic policies of ministries involved in biotechnology. Each year a White Paper on Science and Technology is published as a review of Japan's research and development activities. For instance in 1986, the Science Council proposed to the Minister of Education the promotion of bioscience research at universities and other institutes of higher education.

The policy proposed by the Council for Science and Technology will be followed whilst all aspects of future activities related to the life sciences (biotechnology or biosciences depending upon the ministry) are under review. The Science and Technology agency, which coordinates funding for the promotion of science and technology, is supporting both basic and leading edge life science research. In addition to work at the Institute of Physical and Chemical Research (RIKEN) on life science projects, various aspects of research related to genes, homeostasis, thought processing mechanisms and the preservation of cells, genes and microorganisms are also being supported. The Research and Development Corporation of Japan is supporting the *System for Exploratory Research for Advanced Technology*, as well as contracting activities for the commercialisation of new technologies related to the life sciences. At the National Institute of Radiological Science the effects of radiation of living organisms and cancer therapy using radiation is being studied. At the Japan Atomic Energy Research Institute life sciences research such as the test manufacturing of radioisotopes for medical use is being undertaken. The Environment Agency is supporting various areas of research at the National Institute for Environmental Studies as well as conducting experimental research aimed at pollution prevention. Specific areas of emphasis are as follows:

The Ministry of Education which has the general aim of promoting science, is supporting a large number of projects. Any suitable project may be supported as no specific technical areas are stressed.

The Ministry of Health and Welfare through national research institutes such as the National Institute of Health and the National Cancer Centre, are promoting biotechnology-related research in areas aimed at improving health and the treatment of disease. This includes promotion of research for the prevention, diagnosis and treatment of various diseases through subsidies and commissioned work.

The Ministry of Agriculture, Forestry and Fisheries (MAFF) is attempting to strengthen the country's R & D activities with the aim of ensuring a stable food supply and increasing the productivity of the agriculture, forestry, fishing and food industries. MAFF is also active in the private sector promoting the development of biotechnology and other advanced technologies.

The Ministry of International Trade and Industry is promoting the development of biotechnology at various research institutes established by the Agency of Industrial Science and Technology (a part of MITI). It is also promoting the bioindustry in general by supporting "Research and Development Project of Basic Technology for a Future Industries" as well as the development of medical and other equipment and apparatus using advanced technology. It also harnesses the development strength of private industry through the New Energy and Industrial Technology Development Organisation (formerly the New Energy Development Organisation - NEDO) and promotes R & D in energy conversion technology based on biotechnology.

The Ministry of Construction is promoting the development of new waste water treatment systems which use biotechnology in order to reduce costs and conserve energy.

Other organisations are also dealing with biotechnology include: the Council for Science and Technology of the Prime Minister's Office, the Science Council of the Ministry of Education and the Industrial Technology Council of the Ministry of International Trade and Industry, The Bio-orientated Technology Research Advancement Institution, the Adverse Drug Suffering Relief and Research Promotion Fund and the Japan Key Technology Centre.

The average population of Japan is increasing in line with life expectancy. Hence, medical treatment related to the elderly are seen as particularly important. This requires the establishment of preventive, diagnostic and therapeutic methods for the most common causes of death at the present time, e.g., cancer, cerebral, vascular and heart disease.

A second priority area concerns global environmental problems. In the chemical industry the emphasis is on energy saving and environment friendly processes as well as the utilization of waste products as raw materials. In the area of environmental preservation, the development of bacteria which can efficiently breakdown and deal with harmful substances is also being investigated. In the areas of pharmaceuticals and agricultural chemicals, the emphasis is on development of efficient and manufacturing methods for physiologically active substances derived from biological sources. The food industry already produces many biotechnology dependent products such as isomerized sugars. Future developments are expected in the areas of new flavours and health foods as well as in adoption of techniques which lead to cost reduction and foods ingredients with new functional properties. In the field of agriculture, new seeds and seedlings developed using tissue culture will result in plants which are uniform and of better quality than those derived from seeds. The objective is to increase food production, whilst developing new breeds and varieties as well as artificial seeds (hybrids).

However, activities are not limited to these areas. It is hoped that a greater understanding of the basics of biomimetic techniques will result in a rethinking of the present industrial technology system. In this area most endeavours are still at the planning and research stage. However, there are examples of projects, such as the Human Frontier Science Program, which have already been started whilst a few have reached practical application stages; biosensors being one example. A related topic of particular interest is that of biochips; a logic -chip which uses specific molecular properties of proteins and functions in a manner different from that of semiconductors. Efforts are also being made in the area of protein engineering, a technology which involves elucidation of the correlation between protein structure and function. It is hoped that the accumulation of such basic information will enable the production of proteins which possess the desired function based on a theoretical design. If this is possible and the end results can be effectively applied, the impact on a wide range of industrial activities will be significant.

Interest in biotechnology increased in the early 1980s as results of recombinant DNA experiments became known. This resulted in the establishment of experimental facilities for genetic engineering in Japan. Since 1985 universities, faculties, departments and graduate programs have been either reorganized or newly established in order to improve education related to biotechnology. In 1986 the Science Council submitted a proposal to the Minister of Education entitled "The Promotion of Biosciences Research at Universities". The proposal aimed to promote research activities and development of human resources. It was felt to be extremely important to optimize teaching in order to foster creativity in young scientists taking account of the interdisciplinary nature of biotechnology. To do this it was necessary to revise the framework of science within faculties and universities so that cooperative research could take place. The document also proposed the completion and upgrading of relevant research organisations as well as the reorganization of leading departments, chairs and sections. The report also recognised that within the biosciences there are areas of advanced technology which will require many people to act in support role. These include areas such as analysis and the maintenance and breeding of experimental organisms. Hence, importance was placed on policies aimed at ensuring an adequate supply of technicians.

Researcher from the national research institutes, through a programme set up by the Science and Technology Agency (STA), are sent to national and public universities within Japan, as well as overseas for extended periods of time. The Ministry of Agriculture, Forestry and Fisheries carries out group training of researchers from industry, as well as prefectural and local institutions, at its research facilities. In addition companies provide opportunities for training through joint or collaborative R & D projects with other industries, universities or national research institutes. Cooperative research which involve personnel exchanges are also frequent as a means of further educating researchers from private companies. In Japan there are a large number of scientific societies active in fields related to biotechnology. They publish journals and hold symposia, etc.. Together with relevant international societies and publications this also contributes to improving the quality of research and expanding the knowledge of biotechnology.

The Science Council first prepared safety guidelines in 1979. These were entitled "Guideline for rDNA Experiments in Research Organisations such as Universities" and concerned DNA technology. The Council for Science and Technology then drew up "Guidelines for Recombinant DNA Experiments". Each of these guidelines reflected the most up-to-date scientific knowledge available and went through many revisions before reaching their present form. Investigations were begun in 1983, at the OECD/CSTP, concerning measures for ensuring safety at the commercialization stage, recognising that harmonious international regulations were indispensable in order to prevent international friction and enable trade and investment in the bioindustries. As a result, in July 1986 a report entitled "Recommendation of the Council-Concerning Safety Considerations for Application of Recombinant DNA Organisms in Industry, included the following points: 1) guidelines for proper use within a factory or plant, 2) promotion of further research and information exchange, 3) future studies and research related to closed system guidelines.

At present in each country which received these recommendations, guidelines concerning the safe use of recombinant organisms within plants and factories, have been prepared and made public. In June 1986, the Ministry of International Trade and Industry prepared and made public "Guideline for Industrial Application of Recombinant DNA Technology", while in December of the same year, the Ministry of Health and Welfare made public "Guidelines for the Manufacture of Drugs Using Recombinant DNA Technology". In February 1989 the Environment Agency published its views on release of GMOs in a document entitled "Basic Views on Environmental Safety when Using Recombinant in an open system". The Ministry of Agriculture, Forestry and Fisheries brought into effect the first guidelines, "Guideline for the Utilization of Recombinant in the Agriculture, Forestry, Fishery and Food Industries" in Japan dealing with the practical application or recombinant in open systems in April 1989. In December 1988 the Science and Technology Agency published "Views Concerning Experiments Using Plants", in which standards concerning non-contained experiments using plants were described.

In 1975, the patent system for substances was introduced into the patent system by revision of the patent law, the result of which was that patents on living organisms became recognised. Since 1968, the Fermentation Research Institute has been acting as a deposit and distribution agency for microorganisms for which a patent application has been filled. In 1980 Japan joined the Budapest Treaty, and in 1981 the Fermentation Research Institute was designated by the Director-General of the Patent Office to be the international depository in Japan. In addition to accepting fungi, yeast, bacteria and Actinomycetes for which a patent application was filled, in October 1987 the institution began accepting animal and plant cells, as well as recombinant requiring P2 level physical containment according to the "Guidelines for rDNA Experiments". In 1978, a new law differing from the patent law, entitled "The Seed and Seedling Law" was enacted which related to the protection of new plant varieties, specifically seeds and seedlings for agricultural use. Japan joined the UPOV in 1982. In 1987 the Economic Planning Agency published its "Survey Report on Intellectual Property Rights and Technology Innovation and Transfer". In this report, it was stated that the establishment of an international framework for the protection of intellectual property rights, which accompanies the development of new technological innovation and international development of economic activity, is necessary.

Japan is also viewing the flow of technological innovation and technology transfer from a global perspective, and feels that a new framework for international harmonization of intellectual property rights which contributes to the development of the world economy is necessary. The basic points are as follows; 1) the protection of intellectual property rights which satisfy the distinctive characteristics of new technology, 2) consideration of the effects on society and industry when rights are established, 3) promotion of competition through standardization and application of the Anti-monopoly Act, 4) consideration towards international harmony and developing countries. Industrial activity in fields related to biological phenomena have been expanding due to the development of biotechnology, in particular, gene manipulation techniques such as recombinant DNA and cell fusion. The number of areas of industry involved in biotechnology, for example, pharmaceuticals, foods, main products, chemicals, agricultural chemicals, energy, as well as electronics and machinery, continues to increase. Due to the fact that biological systems are extremely couples and their descriptions difficult, and because in breeding artificial screening and mutations are frequent and reproducibility in insufficient, there have been cases of protection of biotechnology property rights, particularly new varieties of plants and animals (inventions), which cannot be adapted to the patent system. Recently, however, due to rapid technological development, the possibility has been increased.

There is a possibility of new legal measures in the future, due to further development in biotechnology. Development of the bioindustry has been based on a well developed fermentation industry which uses microorganisms to products alcohols, amino acids, organic acids and antibiotics. The experience gained by the fermentation industry became and important base for the introduction and development of "new biotechnology", such as recombinant DNA and cell fusion techniques. In other fields as well, such as plants, Japan has expertised in tissue culture methods, and so on. During the process of application of "new biotechnology", these techniques became integrated with existing know-how gained through years of experience. As a result of this integration, the conventional techniques also underwent improvement as well as becoming more and more widespread. In addition, for example, in the case of biosensors, it can be seen that combination with other fields, such as advanced materials and electronics, is coming about. "New biotechnology" has had a very significant impact on Japanese industry. Industry usually pays close attention to changes in the industrial structure as well as economic and social trends, and attempts to address these changes in an appropriate manner. Enterprises must also evolve as changes occur. However, prior to achieving an improvement in the structure of a company, the diversification of activities is thought to be necessary. Various fields such as pharmaceuticals and health, agriculture and foods, the environment, resources and energy are included in the bioindustry. If we examine biotechnology according to the type of industry involved, the well known industries such as agriculture, forestry and fisheries, foods, chemicals, household detergents, cosmetics, pharmaceuticals. Resources and energy can be mentioned. However, the existence of various other industries which supply analytical and measurement equipment and apparatuses for research purposes should not be ignored.

According to a survey conducted by BIDEC, a wide variety of products will be commercialized by companies in Japan. These include amino acids, cosmetic ingredients, industrial enzymes, physiologically - active substances, pharmaceuticals such as antitumor agents, interferon, thrombolytic agents, interleukins and EPO, as well as monoclonal antibodies, seedlings, and microbial pesticides. The enterprises which will play a major role in the development of these products will belong to the pharmaceutical, chemical and food industries. For the development of a means of support such as analytical and measurement equipment and facilities for research and production, the companies in fields other than the above mentioned 3 fields will play a major role in this area.

Research which is close to commercial development or marketing is, in principle, being carried out by funding of the private sector. The Small Business Finance Corporation has a loan system for funding facilities and operations in order to promote the industrialization of advanced technologies, including biotechnology. There are no mutual interest relationships of any sort between public employees who work at research institutions and biotechnology venture firms.

Intellectual property rights is an important area of international cooperation. Japan is participating in various studies and activities being conducted at WIPO, examples being the international harmonization of patent systems and studies on appropriate protection for biotechnology patents. Japan is also providing technical support to developing countries within the framework of WIPO. Top level conferences and experts committee meetings which are examining common problems between the patent offices of Japan, the United States and European Community are being held. The main themes of these sessions are patent information problems and harmonization of patent systems and operations. Japan is a member of the Budapest Treaty which deals with the deposition of patented microorganisms. The Fermentation Research Institute has been designated as an international depository. Japan has joined UPOV, an organisation involved in the protection of new plant varieties. Japan is a participant in the OCED Committee for Science and Technology Policy (CSTP) expert committee meetings on biotechnology safety issues. Japanese guidelines concerning the industrial use of recombinant DNA were drawn up based on the 1986 OECD Council Recommendations.

Biotechnology, together with electronics and new materials, is believed to be one of the key technologies which will play an important role in improving the welfare of people and improving the future industrial structure. Expansion of basic research, fusion with other areas of science and technology, the promotion of international cooperation and obtaining the understanding and consent of the public are all areas thought to be extremely important for the future development of biotechnology. However, it is worth pointing out here that biotechnology will not merely be useful with respect to the production of substances, but rather, it is also being clearly recognised that as a means of research, it will possesses many more "unknown" areas which have yet to be studied compared to, of course, the land environment, so the future development of marine biotechnology, which focuses on marine organism resources, is being keenly awaited. It is also anticipated that biotechnology will make significant contributions to preservation of the global environment, a topic which lately has become increasingly urgent. This will include the development of environment-friendly products and processes, not only the breakdown and removal of pollutants and dangerous substances from the environment.

(5) NEPAL

Biotechnology is in its infancy in Nepal and no specific National Policy for biotechnology has been formulated. Recently the Government approved the National Science Policy for Nepal in which RONAST plays a leading role in policy formation and is planning to form a national board on biotechnology.

RONAST will be the administrative body and will co-ordinate the educational and research activities in the country.

The most important areas of biotechnology for Nepal are those which have potential use in agriculture (plant and animal); food and beverages; animal feeds; fuels; material recycling; and conservation of genetic resources.

Tribhuvan University is the only University in the country dealing with science and technology. The University has a degree programme in different fields such as biology, zoology, microbiology, medicine and forestry; mostly at BSc level. Higher education opportunities at post graduate level are not available in the area of biotechnology.

An bank for the collection of type culture of differences organisms has been set up by RONAST.

A few alcohol fermenting industries exist. As far as technology applications are concerned all industries are traditional and do not use genetically manipulated organisms and are not collaborating with multi-internationals.

The general public are not aware of new biotechnological development. No effort has been made to promote public awareness. Most scientists are not against new biotechnology but all agree that an infrastructure system should be developed first before initiating research in new biotechnology.

Nepal is looking forward to collaborative work with International Biotechnology Centres/Institutions.

(6) PHILIPPINES

In 1990, the Science and-Technology Co-ordinating Council approved the National Biotechnology Action Plan to Provide R & D support for establishing bioindustries and raising food self-sufficiency. Six mega projects were identified for implementation in the period 1991-1995. Penicillin production using locally available raw materials, Human, animal and plant diagnostics and vaccines, Coconut tissue culture. Coconut tailored fats, Urban wastes and Bioforestation.

The National Institute of Biotechnology and Applied Microbiology (BIOTECH) established at the University of the Philippines at Los Bonos (UPLB) is the most important institution in biotechnology in the Philippines. After one decade of operation BIOTECH reports that it is commercially producing its first research results. In order to further enhance the interaction between BIOTECH and the industry, a 22 hectare Science Park was established at UPLB, in July 1992, with the financial assistance of the Department of Science and Technology. The Science and Technology. The Science Park will commercially be managed by the UPLB Bioindustries Foundation and is intended to attract tenants from the Philippines and the ASEAN region.

Tissue culture research is mainly carried out at university laboratories with governmental funds. Some multinational companies, such as Pioneer Hi-Bred and Del Monte, have established their own research facilities. Research has been done on clonal propagation of coconut, micropropagation of bananas in vitro by using cytokinins to induce proliferation, micropropagation by shoot culture and improved virus elimination techniques in potato, and other research concerns abaca (Manila hemp), citrus, orchids, forest, and root crops.

The main commercial interest in tissue culture techniques is directed at orchid propagation. The majority of the (semi) commercial tissue culture laboratories (22 out of 28) are working solely on orchid propagation. Only five laboratories, three of which are governmentally owned, are working on food crops. These include: abaca, macapuno coconut, banana and pineapple and, to a minor extent, (white) potato.

Commercialization of tissue culture of food crops is hindered by high production costs. Nearly all supplies (such as laboratory glassware, chemicals and equipment) must be imported. The price which farmers have to pay for propagules is consequently high. Purebearing macapuno trees, for instance, can only be propagated by tissue culture but cost US$ 25 each. Only farmers with sufficient financial resources benefit from micropropagation. Small farmers interest in micropropagation is very limited.

(7) SINGAPORE

An emphasis on technology development in Singapore has led to strategic alliances, joint ventures, and a variety of technology transfer arrangements and opportunities. The Singapore Economic Development Board (EDB), a government agency with the goal of enhancing the private sector, has targeted biotechnology and high technology as two prime emerging industry sectors for the 21st century, and has accordingly committed generous funding to local efforts in biotechnology. Since the late 1980s, the EDB has provided over $100 million to the sector through funding the development of the Institute of Molecular and Cell Biology (IMCB).

Focusing primarily on basic research, IMCB is conducting a wide range of studies on cell regulation, neurology, the production of recombinant proteins, DNA/peptide synthesis and sequencing, and HIV. IMCB has strong patent positions in specific areas of cancer diagnostics and treatments, its researchers have published widely, and it is viewed by many in the scientific community as the premier research institute in the Far East.

Areas of joint venture research and product collaborations include molecular diagnostics, combinatorial antibodies, and drugs for colorectal cancer, infectious diseases (including vaccines), AIDS, and compounds based on microbes that are indigenous to equatorial climates. IMCB increasingly enjoys the reputation of being an aggressive purveyor of new biotechnology development. It is currently positioning itself to spin off select technologies into private companies, and has begun the process with the founding of Gene Singapore Ltd. (Genesing), which is focused on developing therapeutics, especially for the East Asian medical market.

In 1990, Singapore's Economic Development Board also established Singapore Bio-Innovations (SBI), which is charged with commercializing local biotechnology developments and locating attractive overseas companies willing to transfer technology to Singapore in exchange for equity investment.

SBI's $20 million fund has provided investment capital to select biotechnology companies from the Untied States, Europe, and Asia.

SBI offices are strategically located in six cities throughout the United States that are considered centers of biotechnology, with a staff that pays close attention to breakthrough research at research institutes and biotechnology companies. It funds projects at levels of $250,000 to $2 million, actively seeking out investment opportunities that will directly impact the more pressing needs of Southeast Asia-treatments for infectious and viral diseases, such as hepatitis B vaccines and therapeutics for cancer, for example.

While therapeutics have an important role to play in technology development in Singapore, basic medical devices, disposable, and diagnostic tools are also important growth sectors. Since the 1990s, government incentives have been initiated to encourage alliances between local companies and foreign firms willing to transfer technology for these products into local companies-with the goal of establishing manufacturing facilities in Singapore and furthering local research and development. Special incentives such as grants, financing, and technology incubation facilities are made available to foreign companies through the National Science and Technology Board.

(8) TAIWAN

Biotechnology continues to play an important role in the Republic of China's six-year national development plan (1991-1996), and has been targeted to receive generous public and private investment. Investment of private capi tal has not necessarily followed government endorsement of biotechnology in Taiwan, due partly to the area's emphasis on computer and other electronic industries.

Efforts are, however, underway to develop a local biotechnology sector; these have been led to Taiwan's Industrial Technology Research Institute (ITRI). The primary biotechnology effort was launched in 1986 with the founding by ITRI of the Development Center for Biotechnology (DCB), which is located at National Taiwan University in Taipei. Within the DCB facility there exists a pilot plant, laboratories, and administrative offices. DCB is focusing on assisting local pharmaceutical companies in basic research efforts.

In 1994, ITRI announced its intention to fund six types of medical research deemed critical to the health needs of the local population: therapeutics and diagnostics for cancer; treatments for cardiovascular disease; products for incontinence; therapeutics for the control of contagious diseases and gastrointestinal diseases; and treatments for neurological disorders of the aging population, such as Alzheimer's disease.

A model for possible alliances with Taiwan might include collaborative research with the Development Center for Biotechnology, or with one of the large pharmaceutical companies. In Taiwan, there is also the Biotechnology Development Association, whose mandate includes establishing an amino acid development task force, promoting effective treatments for bioindustrial wastes (a growing problem in Asia), and establishing links between Taiwan's scientific community and other countries developing advanced bioindustries.

Currently, Taiwan has a handful of biotechnology and pharmaceutical companies located at Hsinchu Science Industrial Park, some fifty miles from Taipei. These companies are involved in various levels of basic research and pharmaceutical production. Lifeguard Pharmaceutical is developing a hepatitis B vaccine for the Taiwan government, and its facility was constructed in the mid-1980s according to good manufacturing practice regulations. General Biological Corporation (also located in Hsinchu) is developing diagnostic reagents. Other areas of interest to the scientific community in Taiwan include molecular biology, cell biology and immunology, biochemistry, and biological process development.

(9) THAILAND

The Ministry of Science, Technology and Energy (MOSTE) works in collaboration with the National Economic and Social Development Board (NESDB) in formulation of plans for science and technology in which biotechnology forms a part. Considerations of relative advantage in terms of resources and demand, and indigenous strength, led to selection of priority areas and targeting of goals in the medium term. Biotechnology is a major concern of the Sixth National Economic and Social Development Plan which also covers science and technology in other sections dealing with aspects such as the development of the production system, marketing and employment. The establishment of the National Centre for Genetic Engineering and Biotechnology (NCGEB) in 1983 represented the first attempt to support a priority area of R&D in a coordinated manner. The Science and Technology Development Board (STDB), established through US-Thai cooperation in 1985, further provides substantial input into priority R&D areas including biotechnology.

National policy on biotechnology is the responsibility of the National Centre for Genetic Engineering and Biotechnology, formed under the Ministry of Science, Technology and Energy. This is to be the main policy, support and coordination centre for biotechnology research and development projects and will form linkage with the private sector. The emphasis is on development, transfer and utilization of biotechnology, including genetic engineering in the following areas: industrial applications, agricultural applications, public health, energy and environmental applications, and strengthening of the basic infrastructure in genetic engineering and biotechnology.

The Science and Technology Development Board, established in 1985 with cooperation of the US Agency for International Development (AID), is another major supporter of biotechnology research and development in Thailand. STDB has special interest in promoting industrially relevant research and development, and has a special programme of low interest loans to industries for company directed research and development. The National Research Council of Thailand sets policy and also plays a supportive role by giving research funding in overall areas of science and technology. Research and development in biotechnology are presently carried out mainly in universities.

The present research strengths are in biomedical and agricultural sciences, while such areas as bioprocess engineering are still to be further developed. Mahiodol University has various active research groups, with emphasis on molecular biology and genetic engineering in various biomedical areas, and especially in those concerning tropical diseases and industrial microbiology. Kasetsart University is active in various areas of agricultural and industrial biotechnology, including plant biotechnology and tissue culture for improvement and propagation of various plants, embryo transfer technology and fermentation technology.

Chulalongkorn University is well known for research on aquaculture of shrimps, rice tissue culture and processing technology. King Mongkut's Institute of Technology Thonburi has various activities on bioengineering and biomass utilization. All of these institutes are located in the Bangkok area. Among the provincial universities, Chian Mai University is the most active, with work on tissue culture of various crops.

Among the government agencies, the Department of Agriculture has a very large research programme, much of which is related to biotechnology.

Thailand is a strong agricultural country presently undergoing transition towards industrialisation. The areas of biotechnology of greatest immediate interest are therefore those concerned with upgrading the quality or increasing the yields of primary products. However, there is increasing interest in biotechnology for processing of agricultural products, and a longer term interest for production of pharmaceuticals and other speciality chemicals. It is noteworthy that none of the commercial firms at present are using genetic engineering, hybridoma or other new technologies in their production at present. The capability in these new areas of technology are by far the predominant concern.

The present education and training programmes still concentrate mainly on traditional areas of science, agriculture and medicine, with little capacity for production of manpower in new biotechnology areas.

A number of regulatory agencies are concerned with standards, safety and environmental impacts of biotechnology-based industries. The Thai Industrial Standards Institute issues both compulsory and voluntary specifications for standard industrial products, including biotechnology products. The environmental effects of industries are monitored by the National Environment Board, which is responsible for enforcement of the Environment Control Law. Although no new laws have been contemplated for regulations concerning biotechnology in particular, the issues concerning laboratory experiments, field trials, environmental release and consumers protection are being studied with the view of setting up appropriate guidelines. The National Centre for Genetic Engineering and Biotechnology have commissioned studies on these topics, and will have an active consulting role in any new regulatory procedure to be set up concerning biotechnology.

As the patenting system in industrially developed countries is rapidly undergoing adjustment with respect to protection of biotechnological inventions, a striking contrast is seen with Thailand, where the first patent law was passed in 1979. The present law in Thailand, not surprisingly, reflects the status of industrial development. The law does not specifically accommodate protection of biotechnological inventions, with which Thailand is still unfamiliar.

Thailand Institute for Scientific and Technological Research (TISTR), the main industrial research centre of the country, has the Microbiological Resources Centre, originally set up with UNESCO support, but now supported by the National Centre for Genetic Engineering and Biotechnology (NCGEB). This Centre performs both service and repository functions. TISTR, Kasetsart University and Department of Agriculture have germplasm banks for various plant species, especially those of economic importance.

The Thailand Development Research Institute (TDRI)surveyed the technological capabilities of 42 biotechnology-based Thai companies in aquaculture, feed, seed, dairy, organic acids, alcohol, and health related industries. The survey identified the following constraints on biotechnology development: inadequacy of science and technology manpower, lack of company interest in R&D, inaccessibility of technological information, weak technical consultancy services and weak private-public sector linkages.

The disappointing state-of-the art of its biotechnology industry, has stimulated the Thai government to take new initiatives. The biotechnological activities are considered to be a core sector in the Thai economy as it will create linkages between the development of the agricultural and the industrial sector.

In December 1991, the *National Science and Technology Development Agency* (NSTDA) was founded as an umbrella organization of among others the *National Centre for Genetic Engineering and Biotechnology* (NCGEB) and the *Science and Technology Development Board* (STDB). The NSTDA is governed by a Board of Directors, including representatives of the private sector and government in equal numbers, and chaired by the minister of Science, Technology and Environment. Government interest in biotechnology has recently also led to the establishment of the *Bioservice Unit* (BSU) as a core facility to provide researchers from both governmental and private organizations with appropriate instruments to strengthen their research capabilities.

The Thai government pursues enhancement of private biotechnology R&D by stimulating foreign capital investment and technology transfer. Governmental incentives focus on value-added export investments. In 1988, 1,400 new foreign investment projects were approved with a total value of US$7.5 billion. Most of this foreign capital comes from Japan and the Asian Newly Industrialized Countries. American biotechnology companies have invested relatively little in Thailand. This is reportedly due to the weak financial position of many US biotechnology firms, the decreasing interest of US financial institutions in biotechnology, and the disagreement about intellectual property rights. To win over the US companies, the *US-Thailand Commercialization of Science and Technology Program* (UST/COST) was launched in July 1990. It involved an unusual set partners: the US *Agency for International Development* (USAID). STDB, *the Maryland International Division* (MID), *the Board of Science and Technology for International Development* (BOSTID) of the US National Research Council, and *Biotechnology International* (BIT), a programme at the University of Maryland. The project also depends on the co-operation of private companies both in Thailand and the USA. The pilot project comprised two distinct stages: 1) an assessment of the needs of Thai industry and agriculture, and of the Thai infrastructure in biological sciences, and 2) the identification and recruitment of Maryland biotechnology companies that could provide the appropriate products and technologies. The first phase was carried out between July and October 1990. The biotechnology needs formulated by Thai companies, concerned the areas of plant production livestock and aquiculture. Demand exists in these fields for new products that will improve the productivity of existing business, such as the production of tomato paste and potatoes. Genetic engineering was expected to be the most suitable technology to reach the desired improvements. In the second phase, it became clear that most of the products and processes required by Thai firms were already available at US biotechnology companies. However, the transfer of such technology faces two problems. Firstly, products and processes need climatological and scale adaptations, especially in the case of agriculture and livestock. The cost of these adaptations has prevented an agreement so far. The second problem concerned the suspension of US foreign assistance to Thailand as a result of the military coup. The project was restarted in late Spring 1991.

Currently, there are over twenty agreements under consideration under the UST/COST programme, with an estimated value of US$ 5 million.

Among the developing countries, particularly in ASEAN region, Thailand has a competitive edge. However, the relationship has not developed into open competition or conflict so far. Instead, there is a national policy of regional cooperation. In the biotechnology field Thailand has collaborative research ventures with the US, Japan and UK including exchange programmes through the JSPS (Japanese Society for Promotion of Science) and the British Council. In addition to such exchange programmes and collaborative research activities the US has also offered grants and loans to Thailand to help develop expertise in biotechnology. It has enabled Thailand to establish the STDB. Thailand also has a good relationship with Australia through a collaborative agreement in business opportunities and development which includes a collaborative R&D project with ASEAN on biotechnology.

The constraints have been in funding, manpower and facilities, but with the development education in biotechnology in recent years and together with the availability of substantial grants from NCGEB and STDB the situation has improved. However, in terms of international comparison, it is still relatively weak, but with the government scheme to offer 180 scholarships for students to study biotechnology abroad in the next three years, and with improvement of education in Thailand, the problem of manpower will be drastically reduced. Funding and facilities problems will cease when STDB and the NCGEB become integrated into a new funding-research agency as planned in 1992.

(10) TURKEY

Strategically located at the border of the European and Asian continent, Turkey may perform an important regional function in plant biotechnology. A strong scientific base in tissue culture techniques has already been built up. The genetic engineering of plants will be initiated soon.

Biotechnology was put high on the national agenda in the late 1980s. In the government's current *Five-Year Development Plan* (1990-1994), biotechnology is included among the priority technology areas. There is no specific national stimulation programme or government budget for biotechnology research, but investments in biotechnology research are encouraged by incentive measures such as tax reductions and import duty exemptions since November 1990. In 1990, the Turkish Bank of Agriculture - a state-owned bank which is the major source of funding for agriculture - opened the *Special Agricultural Projects Department* which aims at financing biotechnology projects.

Before 1990, advanced research was conducted mostly under projects funded by international organizations, such as the UNESCO, the COST Project (*Cooperation European dans la domaine de la Recherche Scientifique et Technique*) of the European Community and NATO, through its *Science for Stability Programme*. Turkey joined the *Biotechnology*

Research Centres Network of UNIDO in 1989. In the same year, the country jcined the *European Network of Biotechnological Applications*, which is funded by UNDP. Its two main branches are (1) microbial culture collections and (2) the exchange of knowledge between Albania, Hungary, Malta, Portugal, Yugoslavia and Turkey, in the fields of agricultural production, chemical products and food additives, aquaculture, sewage management, bioengineering and education.

Advanced education is mainly provided by the universities. Several Biotechnology Units were established in the late 1980s, with full masters and doctorate level courses. Biotechnology research is now undertaken at more than twenty laboratories, at the universities, at the *Turkish Science and Technology Research Institute* (TUBITAK, the major research institute in the country) and at the specialized research institutes of the Ministry of Agriculture.

The total number of researchers currently working in biotechnology is about 300, around 200 are working in plant biotechnology. These researchers are mainly working on tissue culture techniques. Research on the genetic manipulation of plants has recently been initiated.

(11) VIETNAM

Biotechnology is being considered as one of the five top science and technology priorities in Vietnam. Since the early 1980s, there have been two national biotechnology programmes. A new programme for 1991-1995 is currently under way.

Biotechnology research in Vietnam may benefit from the *doi moi* a restructuring policy adopted in 1987 by the Vietnamese government to make the country more self-sufficient. As a part of this policy, Vietnam has opened its markets and laboratories to foreign partners. One of the results is that more and more Vietnamese biologists are returning to their country. Research in Vietnam, however, encounters several constraints. Firstly, governmental support is very limited, impelling scientists to create alternative means to supplement their funding. They have to turn themselves into entrepreneurs to support their institutions, their work and families, many Vietnamese scientists started to look for opportunities to co-operate with foreign laboratories. Secondly, co-operation efforts with the former Soviet Union and Eastern Europe are decreasing. Thirdly, the economic embargo, imposed in 1975 by the United States has seriously limited opportunities for Vietnam's scientific and technological exchange with foreign countries. The embargo puts political pressure on major American trading partners, such as Japan, to limit business and scientific exchange with Vietnam. Nevertheless, countries like South Korea, Taiwan and Singapore have been investing heavily in Vietnam since the *doi moi* began. The *doi moi* has stimulated agricultural production. Meat production has increased from 33 to 1037 thousand tonnes. Coffee has become one of the most important export corners: since 1986, the amount of exported coffee increased from 19 to over 45 thousand tonnes. Production of rice, however, is not growing fast enough. The former Soviet Union used to supply most of these products on favourable terms, but Vietnam's huge rouble debt prevents any further sale on credit.

In spite of all the constraints, Vietnamese biotechnologists have achieved some promising results. The efforts of the *Biotechnology Centre* (BC) at the *National Centre of Scientific Research* (Hanoi) on disease-free potatoes are wellknown. In the past, the Vietnamese relied heavily on European potato seed. But imports of new seed tubers fell during the Vietnam war, and existing stocks became infected. Fear of shrinking yields prompted the government to encourage scientists to work on tissue culture and regeneration.

The interest in biotechnology in Vietnam is based on the application of biological knowledge, particularly microorganisms and modern techniques. Present interests include traditional processes such as foods (fermented soybean, putrefactive soybean curd, fish sauce and glutinous rice alcohol).

Other traditional aspects include the use of the Azola symobioic association as agree manure in paddy fields and the cultivation of mushrooms. Other topics of interest include modern processes the production of microbial metabolizes such as amino acids, organic acids and antibiotics as well as the cultivation and use of giberellins for improving the crop yield of some kinds of vegetables which has succeeded commercially.

The application of modern techniques such as cell and pollen culture protoplast fusion is of interest in plant breeding of orchids, potatoes, pineapples, sugar canes whilst the cultivation of microalgae as single cell protein, is at pilot scale. The bioconversion of waste and renewable materials is developing on a commercial scale and includes bakers yeast from molasses, glucose syrup from manioc, and biogas from animal excrements.

Basic methods of the DNA recombinant technique have been practiced successfully in laboratories. The purification of restriction enzymes, plamid DNA without using ultracentrifuge and transformation of *E.coli* and *Saccharomyces cerevisiae* has been accomplished. The cloning of cellulose genes is in a preliminary stage.

III. GENERAL CONCLUSIONS

The world has become a global village where information and collaboration are of paramount importance for those who want to be competitive players in research and the world markets. The biotechnology community has a major role to play. It will, among other things, have to act and work in a manner that will compel the government to recognize the crucial role research and biotechnology development as well as the development of people with excellent skills and knowledge will play in ensuring the future prosperity of the country.

This study provides science policy-makers (those who make decisions on the management, organization, directions and funding of research), industrialists (those who make judgements as to which markets to enter and with which products and processes), trade officials (those who manage and/or control international trade), and concerned citizenry (those who worry about the possible social and environmental impacts of biotechnology) with Asian experience in biotechnology developments. The Asian experience as an ideal model for biotechnology development could be useful for the following reasons :

* In preparing its own national biotechnology programme, a country could consider the Asian experience to avoid mistakes and to adopt measures that work.
* By being aware of where the best or the most appropriate biotechnology research and development is being performed in Asia, scientific managers can line up mutually beneficial cooperative research projects and joint ventures between their units and Asian counterparts.
* Having access to timely and more complete information may allow one's industrialists to better compete

on Asian markets and identify new commercial opportunities.

* Information on the state of biotechnology and its likely progress may have security implications.
* The applications of biotechnology may have effect on the environment that transgress national boundaries. Thus, nations that could possibly be affected may feel the need to have access to relevant information.

This monograph examines the principal fields of biotechnology research and development that are being promoted by universities, government institutes and centres in some Asian countries. Biotechnology research in Asia is conducted by numerous research units developing project devoted to health, chemical and environmental sectors, agriculture and agroindustry. Asian countries are presently embarking on the development of a wide spectrum of biotechnology research as a result of several decades of efforts invested in the development of biotechnology as well as well-defined policies for the enhancement of this field of research. The development of a research base in biotechnology has been produced as a response to the demand, and needs of the health, food and environment sectors. The pattern of development is characterized by a vague connection to socio-economic constraints.

Considering the importance biotechnology is acquiring in the international scene, the potentials it offers to overcome agricultural, food, medical, industrial, and environmental problems and given the existence of an important research base in this field in Asian countries, it becomes relevant to examine the Asian experience in biotechnology as an ideal model for biotechnology development. The evaluation of biotechnology development in Asia was based on the application of four criteria: scientific and technical viability; economic feasibility; characteristics of existing political opportunities for the development and fostering of research areas; potential application and utilization of biotechnology for solving the problems that Asian countries face.

In the Southeast Asian countries of Singapore, Hong Kong, and Taiwan, biotechnology has became an important technology sector in the last few years. No longer do these countries simply want to manufacture generic products that have gone off-patent. Today, the goal is basic research. In the United States, the life sciences have had a secure place in the biotechnology arena for over twenty years, but it is only since the late 1980s that the more industrialized Southeast Asian countries have begun to look upon biotechnology as a sector in which they desire to play a role. The key elements necessary to foster this development usually were missing : established research entities and private capital. By 1992, however, important research were near completion in Singapore, Hong Kong and Taiwan, having been financed primarily through government funding. This is typical of countries in East Asia, where the government supports certain emerging industry sectors with generous funding. As this century comes to a close and the 21st century begins, Southeast Asian countries will play increasingly important roles in the sphere of life science and leading edge medical products.

REFERENCES

Acharya, R. (1995). The impact of new technologies on economic growth and trade : A Care study of Distechnology. Masstricht.

Asian and Pacific Development Centre (APDC) 1988). Regional Seminar on "Public Policy Implications of Biotechnology."

Atthasampunna, P., Suwana-adth, M. and Bhumiratana, A. (1988). Current Biotechnological Development in Thailand.

Ayyapan, S. (1988). Biotechnology and Aquiculture. Asian Fisheries Society, Indian Branch Newsletter, October 1988-March 1988, pp.7-8.

Balaram, P. and Ramaseshan, S. (1991). Biotechnology in India. Special issue of Current Science, Vol. 60.

Biotechnology and Development Monitor (1992). Private Investment in Biotechnology Promoted in India. June Issue.

Biotechnology and Development Monitor (1992). Turkey: Biotechnology receives priority status. September issue.

Board on Science and Technology For International development (1982). Priorities in biotechnology research for international development. National Academy Press, Washington, DC.

Bordier, S. (1990). Biotechnology in Vietnam. Bifuture, March issue.

Brenner, C. and Komen, J. (1994). International Initiatives in Biotechnology for Developing, OECD Development Centre, Technical papers, No. 100, Paris.

Brenner, C. (1988) New Technologies for Agriculture: Developing Country Prospects. Development Centre, OECD, Paris, pp.72-23.

Bull, A., Holt, C. and Lilly, D. (1982) Biotechnology International Trends and perspectives. OECD, Paris.

Bunders, J. (1988) Appropriate biotechnology for sustainable agriculture in developing countries. Trends in Biotechnology. 6:173-180.

Chaudhuri, H. (1968) Breeding and Selection of Cultivated Warm-Water Fishes in Asia and the Far East. Fish FAO. Rep. 44, 4: 30-66.

Clark, N. and Juma, C. (1991) Biotechnology for justifiable development : Policy options for Developing Countries. Acts Press, African Centre for Technology Studies, Nairobi.

Coben, J. (1994). Biotechnology Priorities, Planning and Policies : A framework for decision making. ISNAR Research Report No. 6, The hague.

Coleman, R. (1988). National Policies and Programmes in Biotechnology. In : OECD, Biotechnology and the changing role of government. Organisation for economic co-operation and development, Paris, pp.83-94.

Commandeur, P. and Pistorius, R. (1992). Commercial biotechnology gets shape in South East Asia. Biotechnology and Development Monitor 12 : 11-13.

Dibner, M. (1989). Strategic alliances are likely to remain important in Japan's biotech industry. In Venture Japan, Vol. 2 (1) Asia Pacific Communications Inc., San Francisco, pp. 47-52.

Dixon, R. (1988). Forest biotechnology opportunities in developing countries. Journal of developing Areas. 22: 207-218.

FAO (1979). Agriculture Development in China. Rome, FAO.

FAO (1988). Production Year Book, Vol. 41. Food and Agricultural Organisation of the United Nations.

Fishlock, D. (1989). Biotechnology Survey. Financial Times, London, 12 May, 33-35.

Getubig, I., Chopra, V. and Swaminathan, M. (1991). Biotechnology for Asian Agriculture: Public Policy Implications. Asian and Pacific Development Centre, Malaysia.

Gibbons, G. and Yongyuth, Y. (1989). ASEAN Country Studies. World Bank-ISNAR-AIDAB-ACIAR Biotechnology Study Project papers. ISNAR, The hague.

Gozen, A. (1991). Plant Biotechnology in Turkey: Assessment of opportunities and constraints. Master thesis, Research Policy Institute, Lund, Sweden.

Hacking, A. (1986). Economic Aspects of Biotechnology. Cambridge University Press, Cambridge.

Hamilton, J. (1988). A biotech bit and a bold way to profit from it. Business week, 19 September, 58.

Hulse, J. (1985). Biotechnology: Opportunities and Constraints IORCMR lloe, International Development Research Centre, Ottawa.

IRRI (1984). Biotechnology and International Agriculture Research. International Rice Research Institute, Los Baros, Philippines.

Ishii, T. (1990). Science and Technology Research in Japan's Future, Reference Reading Series No. 22, pp.6-15, Foreign Press Centre.

Katter, R. (1985). The new biotech agriculture: Unforeseen economic consequences. Issues in Science and Technology. 2:125-133.

Kenny, M. (1986). Biotechnology: The University -Industrial Complex. Yale University Press, New Haven, CT.

Kim, K. (1986). Status of Korean biotechnology and its role in improving crop production. In : IRRI Workshop on Biotechnology for Crop Improvement: Potentials and limitations. International Rice Research Institute, Los Banos, Philippines.

Komen, J. and Persley, G. (1993). Agricultural Bistechnology in Developing Countries : A Cross-Country Review. ISNAR Research Report No. 2, The huge.

Leodin, A. (1989). National Policy of Biotechnology in Indonesia. The 5th National Congress of Microbiology, Yogyakurta, December 4-6.

McConnell, D. (1985). Genetic Engineering and biotechnology in Egypt. Mission report to UNIDO.

McConnell, D., Doehmer, J., Normarki, S. and Nuti, M. (1985). Technical Assistance for the Egyptian National Centre for Genetic Engineering and Biotechnology. Mission report to UNIDO.

Miller, D. (1996). Biotechnology in Southeast Asia. Nature Biotechnology, 14:960-961.

OECD (1989). Biotechnology-Economic and Wider Impacts. Organization for Economic Co-operation and development, Paris.

OECD (1992). Biotechnology, Agriculture and Food. Paris.

Omar, A. (1986). Policy and strategies for scientific development in Malaysia with special reference to biotechnology. Proc. XII Malays. Biochem. Conference, pp.19.

Persley, G. (1990). Agricultural Biotechnology: Opportunities for International Development. CAB International, Wallingford, UK.

Prentis, S. (1984). Biotechnology a New Industrial Revolution. Oribis, London.

Rosario, A. and Guzman, E. (1982) . The Status of Plant tissue culture in the Philippines. In Rao, A. (ed.) Proceedings of COSTED Symposium on tissue Culture of economically important crops, Singapore, pp. 293-294.

Ruddle, K. and Rondinelli, D. (1983). Transforming Natural Resources for Human Development: A Resource Systems Framework for Development Policy. United Nations University, Tokyo.

Ruttan, V. (1982). Agricultural Research Policy. University of Minnesota Press, Minneapolis.

Sasson, A. (1988). Biotechnologies: Challenges and development. UNESCO/CTA, pp.282-283.

Sasson, A. (1988). Biotechnologies: Challenges and Promises. UNESCO Sextant, 2:254-255.

Sasson, A. (1989). Feeding Tomorrow's World Paris, Unesco, Technical Centre for Agricultural and Rural Cooperation (CTA), PP.780.

Sasson, A. (1993). Biotechnologies in Developing Countries: Present and Future. Vol.1: Regional and National Survey. UNESCO

Satrapradja, D. (1986). National Policy on Biotechnology in Indonesia. In Workshop on Biotechnology in Agriculture. National Academy Press, Washington DC., pp.50-53.

Sawahel, W., Sastry G, Knight G and Cove D (1993). Development of an electro-transformation system for *E.coli* DHIOB. Biotechnology Techniques. 7 : 261 - 267.

Sawahel, W. (June, 1995). Genetic engineering : The double sgeed sword. Science magazine (Egypt) 225 : 31-34.

Sawahel, W. and Ali, M. (1994). Callus induction and maintenance of *Zea mays* kernels. Biotechnology letters. 16 : 397-400.

Sawahel, W. and Cove, D. (1992). Model system for studying plant molecular biology. Biotechnology Education 3 : 167-172.

Sawahel, W. and Fukui K (1995). Gene cloning in plants : Innovative approaches. Bio/Techniques 19 : 106-115.

Sawahel, W. and Peachy L. (1993). The dilemma of biotechnology in the Third World. Biotechnology Education 4:3-8.

Sawahel, W. and Cove, D. (1992). Gene transfer strategies in Plants. Biotechnology Advances. 10 : 393-412.

Sawahel, W. Onde S, Knight G & Cove D (1992). Transfer of foreign DNA into *phuscomitrella patens* using the gene gun. Plant Molecular Biology Reporter. 10 : 314-315.

Sawahel, W. (1993). Gene transfer strategies. Rice Biotechnology quarterly 14 : 35-37.

Sawahel, W. (1994). A proposed post-graduate programme in plant genetic engineering. Biotechnology Education 3:1-7.

Sawahel, W. (1994). Direct gene transfer & rescue systems in plants. Rice Biotechnology Quarterly. 19 : 11-12.

Sawahel, W. (1994). Kernel - derived callus of *Zea mays*. Rice Biotechnology Quarterly. 19 : 25-26

Sawahel, W. (1994). Trangenic Plant Technology. Rice Biotechnology Quarterly 18 : 13-14.

Sawahel, W. (1994). Transgenic Plants : Performance, Release and Containment. World Journal of Microbiology & Biotechnology. 10 : 139-144.

Sawahel, W. (1995). Genetic manipulation using U.V. laser microbeam. Rice Biotechnology Quarterly. 22:23.

Sawahel, W. (1996). Establishment of Arab Centre for genetic engineering and biotechnology : A proposal for action. Al-Ahram Journal 2/8/1996.

Sawahel, W. (1996). Genetic Engineering Revolution. Scientific Advances (Kuwait) Vol. 14: 13-15.

Sawahel, W. (1996). Identification and Molecular Analysis of Rice Somatic Hybrids using Foreign Transgenes. biotechnology Letters. 18 :909-912

Sawahel, W. (1997). Molecular analysis of genomic DNA - mediated transformation. Biologia Plantarium. In press.

Sawahel, W. (1996). Plant Molecular and Cellular Laser Microsurgery. World Journal of Microbiology & Biotechnology. 12;1-5.

Sawahel, W. (1996). Transgenic Strawberry. Rice Biotechnology Quarterly. 26 : 33

Sawahel, W. (1996). Ultrasound-mediated stable transformation of potato tuber discs. Biotechnology Techniques. In press.

Sawahel, W. (August, 1995). DNA finger-print and forensic medicine. Al-arabi magazine (Kuwait) 441 : 83-86.

Sawahel, W. (August, 1995). Test-tube Plants and gene technology. Al-Faisal magazine (Saudi Arabia). 226 : 83-86.

Sawahel, W. (December 1995). Genetic resource's war. Science magazine (Egypt) 231 : 4-5

Sawahel, W. (July, 1995). Gene Bank. Science magazine (Egypt) 26 6-7, 11.

Sawahel, W. (November, 1995). New scientific revolution : gene finger-print discovers difficult crimes easily. Radio & TV. magazine (Egypt). 3164 : 76-77.

Scheidegger, A. (1990). Biotechnology in Japan towards the year 2000 .... and beyond. Trends in Biotechnology, 9: 183-190.

Sudarmono, P. (1986). The development of biotechnology in Indonesia. University of Indonesia Press, Jakarta.

Suwana-adth, M. (1986). Biotechnology in Thailand, Proceedings of the Conference of First ASEAN Science and Technology week, Malaysia. pp. 239-246.

Taylhardat, A. (1989). The ICGEB: a Center of Excellence. Biofuture. 77:45-51.

UNCTAD (1991). Trade and development aspects and implications of new and emerging technologies: The case of biotechnology, prepared for the trade and development board, Committee on transfer of Technology, Geneva.

United Nations Department of Economic and Social development (1992). Biotechnology and development: Expanding the capacity to product food in advanced technology assessment system. Issue 9, New York.

US Office of Technology Assessment (1981). Impacts of Applied Genetics. Washington, DC.

US Office of Technology Assessment (1984). Commercial Biotechnology: An International analysis. Washington, DC

World Bank (1988). World development Report, Oxford University Press, New York.

World Bank (1989). Agricultural Biotechnology Study. Technical Report, World Bank, Washington, DC, December 1989, pp.54.

Yuan, R. and Hsu, M. (1992). Biostrategies in the pacific Rim. Genetic Engineering news, January issue for Asian Agriculture. Kuala Lumpur, APDC.

Yuthavong, Y. and Bhumiratuna, S. (1989). National Programs in Biotechnology for Thailand and other Southeast Asian Countries : Needs and Opportunities. In : Proceedings of the conference on strengthening collaboration in biotechnology: International Agricultural Research and the Private Sector, April, USAID, Washington DC, pp. 33-52.

Yuthavong, Y. (1987). The impact of biotechnology and genetic engineering on development in Thailand. Journal of the Science Society of Thailand. 13 : 1-13.

Zamora, A. and Barba, R. (1990). Status of Tissue Culture Activities and the Prospects of their Commercialization in the Philippines. Australian Journal of Biotechnology. January issue.

INTERNATIONAL SERIES OF GENETIC ENGINEERING AND BIOTECHNOLOGY

1. Biotechnology Policy and Action Plan for Economic Development in the Third World.
2. Biotechnology development : The Asian Experience.
3. Plant Biotechnology Education Programmes : From School to University.
4. Transforming Biomedical Education for the Gene Age.
5. Plant Genetic Engineering : From A to Z.
6. Plant Genetic Transformation Technology.